SOIL

by Robin Nelson

first step nonfiction

Lerner Publications Company · Minneapolis

We live on **Earth.**

Earth is made of
different things.

Earth is made of water,
gases, rocks, and soil.

Soil is the top part
of the **ground.**

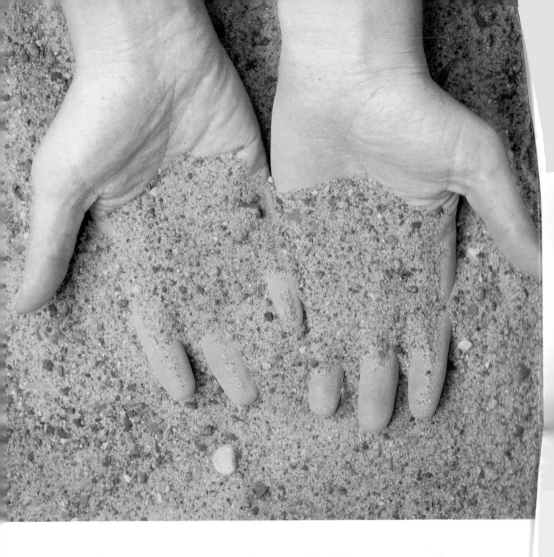

Soil is made of tiny rocks.

Small pieces of dead plants
and animals are in soil.

Soil can be dark black.

Soil can be light brown.

Pieces of soil can be small.

Pieces of soil can be big.

Soil can hold water.

Soil can help plants grow.

We grow flowers in soil.

We grow food in soil.

Worms live in soil.

Soil is found on Earth.

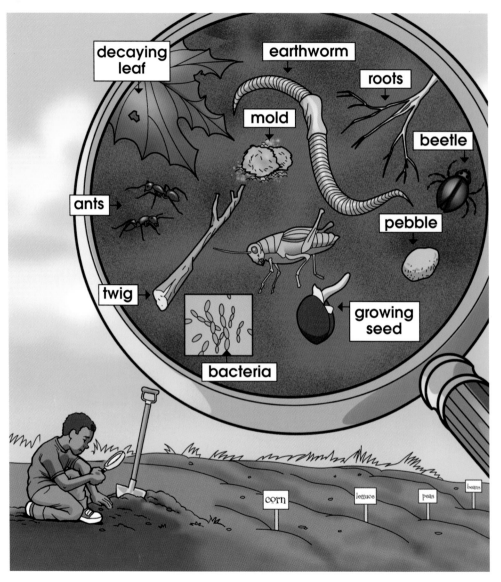

decaying leaf

earthworm

roots

mold

beetle

ants

pebble

twig

growing seed

bacteria

corn

lettuce

peas

beans

18

Soil Community

The ground under your feet is full of life. The plants and animals that live in soil come in many different sizes. Many are too small to see without help. Plants and animals make new soil and keep it healthy. The animals dig in the soil and move it around. They break up the soil when they eat it. Each tiny plant and animal works together in a community.

Super Soil Facts

 It takes hundreds of years to make one inch of soil.

 Almost all the food you eat, material for the clothes you wear, and wood for the house you live in is made using soil.

 One cup of soil holds as many bacteria as there are people on Earth. That's over 6 billion!

 Soils can come in a variety of colors — black, red, yellow, white, brown, and gray.

Soil helps protect the air and control Earth's temperature.

 Some farmers plant rows of trees next to their fields. These windbreaks stop the soil from blowing away.

 Soil is like a sponge. It holds water, which helps plants grow.

 Water, wind, and ice can wear away soil.

Glossary

 Earth – the planet where people live

 ground – the surface of the earth

 soil – the top part of the ground

 worms – small animals with long soft bodies and no legs

Index

animals – 7, 19

Earth – 2, 3, 4, 17, 21

flowers – 14

food – 15, 20

plants – 7, 13, 19, 21

worms – 16

The photographs in this book are reproduced through the courtesy of: © Todd Stand/Independent Picture Service, cover, pp. 6, 8; NASA, p. 2, 22 (top); © ML Sinibaldi/CORBIS, p.3; © Bill Ross/CORBIS, p. 4; © Royalty-Free/CORBIS, pp. 5, 10, 12, 15, 17, 22 (second from top), 22 (second from bottom); © Robert Pickett/CORBIS p. 7; © David Lees/CORBIS p. 9; © Roger Wood/CORBIS, p. 11; © Michael Boys/CORBIS, p. 13; PhotoDisc Royalty Free by Getty Images, p. 14; © Macduff Everton/CORBIS, pp. 16, 22 (bottom).

Lerner Publications Company
a division of Lerner Publishing Group
241 First Avenue North
Minneapolis, MN 55401 U.S.A.

Website address: www.lernerbooks.com

Library of Congress Cataloging-in-Publication Data

Nelson, Robin, 1971–
      Soil / by Robin Nelson.
         p.   cm. — (First step nonfiction)
      Includes index.
      ISBN: 0–8225–2612–3 (lib. bdg. : alk. paper)
      1. Soils—Juvenile literature. 2. Soil biology—Juvenile literature. I. Title. II. Series.
   S591.3.N45 2005
      631.4—dc22   2004020787

Manufactured in the United States of America
1 2 3 4 5 6 – DP – 10 09 08 07 06 05